DA

Discovering
Beekeeping

SHIRE PUBLICATIONS LTD

Contents

Fig. 1. Pollen-carrying worker bee

ACKNOWLEDGEMENTS
The author wishes to acknowledge her debt to the late Frank Vernon, who supplied the photographs and made helpful comments. His cover picture shows bees subdued with smoke.

Published in 1996 by Shire Publications Ltd, Cromwell House, Church Street, Princes Risborough, Buckinghamshire HP27 9AA, UK.
Copyright © 1977 and 1996 by Daphne More. Number 226 in the Discovering series. ISBN 0 7478 0318 8. First published 1977, reprinted with amendments 1988. Second edition 1996.

Printed in Great Britain by CIT Printing Services, Press Buildings, Merlins Bridge, Haverfordwest, Dyfed SA61 1XF.

Introduction

Why keep bees ? Because they make honey, one of the healthiest and most natural foods, which man has sought throughout history, first as a honey-hunter robbing wild bees' nests and later by keeping bees under his own control. This is the obvious answer, but there are others. Bees are fascinating creatures and there is always something new to learn about them. Beekeeping is an enjoyable open-air hobby, bringing contact with kindred spirits from all walks of life. Beekeepers must be nice people: tradition has it that bees do not tolerate the arrogant, the impatient and the immoral nor thrive where there is strife in a household! Again, bees are essential to the national economy; their importance as crop-pollinators far outweighs their value as honey-producers. In built-up areas where wild bees are scarce, due to a lack of rough ground where they can breed, even one hive of honeybees will greatly benefit gardeners and allotment-holders in the district.

Can anyone keep bees? Yes — given the interest and temperament, excepting only a tiny minority who are allergic to bee stings. No great physical strength or courage is needed. Even a garden is unnecessary: hives of bees have been kept successfully on the roofs of buildings in the middle of London, on flat-roofed garages and in lofts. Even in urban areas, gardens and parks can provide ample forage for one or two colonies. However, a beekeeper has a responsibility to manage his hives so that neighbours are not alarmed or inconvenienced. Some old beekeepers cling to the laissez-faire methods which prevailed in more spacious days, allowing swarms to fly off and settle in roofs and chimneys. Apart from the ill-feeling generated, this is uneconomic as bees yield far more honey if swarming is controlled. Hives and equipment are quite expensive, and it is more sensible to have one properly managed and productive colony than three or four neglected ones.

It is impossible to cover the whole subject of beekeeping in this little book. The beginner should find here enough to enable him to set up one or two hives and run them efficiently, and to provide a sound basis if he later wants to expand his beekeeping activities. It must be said that no book can be a substitute for competent personal tuition and demonstration. Readers are advised to join their nearest branch of the British Beekeeper's Association (the General Secretary of the BBKA, see appendix, will provide the address), and also to take advantage of any evening classes or courses run by the county agricultural college (plate 1). In this book I have tried to say *why* things are done as well as *how,* but explanations are necessarily brief. Enthusiasts — and beekeepers are all enthusiasts — will want to pursue the subject in greater depth, so a list of reliable books for further study is included at the end.

3

1. The life of the honeybee

Some knowledge of the honeybee's life is necessary if the beekeeper is to understand the behaviour of his charges. There are other kinds of bees such as solitary bees and bumblebees, but honeybees alone store sufficient honey to allow the whole colony to survive the winter. This means that there are always workers (undeveloped females which do not breed) to do the chores, and the queen honeybee has become a highly specialised egg-producer who does not even feed or clean herself. A colony at full strength in early summer will consist of a queen, up to one thousand drones, and fifty thousand workers. Drones are the male bees and are only present for part of the year.

After the winter break from breeding, the queen (plate 2) begins to lay eggs as early as January, increasing production as early pollen becomes available. The egg hatches on the third day into a tiny larva (plate 3) which for the next three days is fed on brood food or bee milk (also called royal jelly) produced by glands in the heads of worker bees. On the fourth day, worker and drone larvae are switched to a diet of nectar or dilute honey and pollen, but larvae destined to be queens are fed brood food throughout. As the weather improves, workers fly out to fetch early nectar, pollen, and water for diluting the stored honey. By the ninth day, worker larvae are ready to be sealed in their cells under porous wax caps, where they pupate. Twelve days later they gnaw their way out to begin their adult lives.

Drone eggs are laid in March in the larger cells found at the edges of the outer combs of the brood nest, that group of combs where breeding occurs. Drone larvae are sealed beneath distinctive domed caps on the tenth day and the adults emerge on the twenty-fourth. Drones are burly, squarish bees with very large eyes and no sting (plate 4). Their duty is to impregnate the young queens. A drone comes from an unfertilised egg, so has genes only from his mother. Female bees come from fertilised eggs, and it is still not known exactly how the queen always lays the right sort of egg in the different cells.

A queen is both larger and longer than a worker bee and lacks the equipment for collecting and processing pollen and nectar and the glands which produce brood food and wax. She has a sting but only uses it against other queens. Whether a fertilised egg develops into a worker or a queen depends on its subsequent treatment, notably the more lavish food given to queen larvae. Queen cells are used only once: they are large and have a free-hanging position often at the edges of the combs (plate 5). Many of the queen cups seen on the combs are never used, but when the queen lays in one, and conditions are right, the workers extend it to accommodate the fast-growing larva. Should a queen die when there is no egg in a queen cup, the workers will select one or more worker eggs (or failing them, larvae less than three days old) and, tearing down the surrounding cells, will construct an emergency queen cell around each one. These cells will therefore be found on the faces of the combs rather than at the edges (plate 6). Thus the bees can rear a new queen at any time, so long as they have eggs or very young larvae available in worker cells.

The table below sets out the duration of each stage in the development of queens, workers and drones.

	Queen	*Worker*	*Drone*
Days in egg	3	3	3
Days as larva	5	5	6
Days after capping	7	13	15
	15 days	21 days	24 days

There are various reasons why queen cells (between two and thirty) are begun. The queen may be old, dead or diseased and so must be replaced, or the hive may be overcrowded. A queen produces a secretion called queen substance from the mandibular glands behind her jaws; this spreads over her body and is taken from her by the workers who constantly lick and clean her, and is passed to others as they share food. So long as all get sufficient queen substance the bees remain content, but if the queen is dead or failing, or congested conditions in the hive prevent proper distribution, lack of it will trigger off the urge to build queen cells. There is no mind which decides such things: it is an automatic response to circumstances.

The first queen to emerge may find and tear open the other queen cells, kill the old queen if still present, and simply take her place. Some bees regularly supersede their queens in this way, without swarming, and this makes them popular with beekeepers. However, the characteristic tends to disappear, as the new queen may mate with a drone from a strain of inveterate swarmers. Usually the old queen leaves with a swarm, about half the colony, as soon as the first queen cell is capped. Six days later, the first virgin may leave with a cast, about half the remaining bees, and sometimes a second virgin with another. A colony can virtually swarm itself out of existence. Where a queen has died, there is less likelihood of a swarm leaving as the colony will be depleted by the break in breeding, since a queen at her peak may lay two thousand eggs a day. A queen will be at least five days old before she mates and will start laying about five days after that. The sperm she receives at this time lasts her throughout her life.

This timetable can be affected by the weather which may prevent a virgin mating or a swarm leaving although they are ready. A late spring may hold up colony development so that the bees do not swarm at all that year.

The swarm, after swirling about in the air for some minutes, will begin to settle and form a cluster, usually on a tree (plate 11). Scouts will go out to look for a cavity, and as soon as a suitable place is found in a roof or hollow tree the swarm will move in and begin building combs. A swarm gorges itself with honey before leaving the nest so that the bees can survive for a few days, but they must store food as soon as cells are ready because they have no reserves if bad weather comes. A mated queen will begin laying at once, and the workers take up their various duties, the younger bees which have active brood-food glands attending to the larvae and the queen, the older ones waxmaking, and those still older foraging outside the nest. At different stages they are involved with cleaning, ventilation, honey-processing and guard duties. Most workers become foragers about the twelfth day of adult life. They first take a few *playflights* to learn the location of the hive, and as many young bees come out together novice beekeepers often mistake them for a swarm rising. Once established as a forager, a worker will continue so until she dies, probably about three weeks later. Autumn-born bees have a longer life. Emerging into a colony where breeding has stopped, the worker finds no outlet for her brood food and so remains a 'young bee' probably for months, until the presence of larvae prompts her to feed them and she begins to age normally.

Workers may be killed by predators or the beekeeper, by poison or disease, but the majority die of decrepitude, usually away from the hive. A queen may live five years but will be killed by a rival or the workers when fertility diminishes. However, an old queen who leaves with a swarm will often be replaced by the workers before the winter. Drones, their usefulness over, are expelled from the hive to die of cold and starvation. If drones are still present at the end of summer, something is amiss (see page 44).

Foragers collect pollen, which adheres to the hairs on their bodies, is collected by brushes on the bees' legs and is packed into the 'pollen baskets' on the hind legs for transport (plate 7). Nectar is siphoned from the flowers into the bee's honey sac, where the process of converting the sucrose it contains into glucose and fructose is begun. The work is continued in the hive by the younger bees. The sugar content of different nectars varies considerably, but always much moisture must be evaporated to turn it into honey containing less than twenty per cent water. Thousands of bee journeys must be made to collect 3lb (1.362kg) of nectar, which will eventually become 1lb (454g) of honey. Bees only seal it when it has reached the perfect consistency, as too high a water content would allow it to ferment. In addition to what they and the larvae consume and what is left in the brood box for the winter, the bees of one colony may store 100lb (45.5kg) of honey in the supers in one summer. However, to achieve such a crop, weather, available forage and the beekeeper's management would all need to be excellent. 25lb (11.35kg) from a hive is a more realistic estimate of the British beekeeper's average yield.

As breeding declines in autumn, the empty cells in the brood box are filled with nectar from late-flowering plants like ivy. The first cold nights send the bees into a tight cluster on the combs and this is how they spend the winter. They do not hibernate but move slowly over the combs eating the stores. Occasionally, if the temperature is above 10 Centigrade (50 Fahrenheit), they will leave the hive for a *cleansing flight* — to relieve themselves — and then reform the cluster. In the southern counties bees can sometimes be seen on the wing close to their hives in the middle of Christmas Day.

2. Choosing the hive

It is wise to meet bees at close quarters a few times before getting your own. Find out whether you like them before, not after, spending your money. Attend an open day at an agricultural college, or a summer meeting of your local beekeepers' association (you will be welcome even if you are not yet a member) and watch the demonstrator opening a hive. You need a simple bee-hat with a veil (probably you can borrow one). Bees will fly around and may settle on you but keep still and they will not harm you. A zipped-up anorak with tight cuffs and trousers tucked into wellingtons will prevent any bee getting inside your clothes, which might upset you both.

Having decided to keep bees, you must choose a hive. I strongly advise new beekeepers not to consider the gable-roofed WBC, even as a gift. This old-fashioned hive is difficult to operate successfully, even after years of experience, and defects in the design make it almost impossible to avoid upsetting the bees (and incidentally the beekeeper) during routine examinations. Because it has inner and outer walls, there are many extra parts to handle and maintain, and it takes twice as much timber, making it the most expensive hive to buy. Its size and the fixed splayed legs also make storing and transporting it awkward.

Modern hives consist of the following parts:

Floor: A shallow tray with one side open to form an entrance and a block which fits into it to form a small winter entrance or to close it completely.

Brood box: A bottomless box with identical perimeter dimensions, within which removable light wooden frames containing combs are hung by lugs from rebates in two walls.

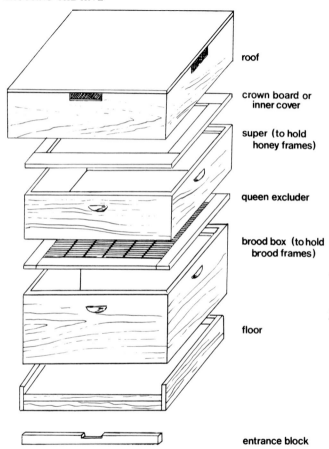

roof

crown board or
inner cover

super (to hold
honey frames)

queen excluder

brood box (to hold
brood frames)

floor

entrance block

Fig. 2. Parts of a typical modern hive.

Queen excluder: A grill which allows workers to pass through but
prevents the queen and drones passing from the brood box
where breeding takes place into the supers which are solely for
honey storage.
Supers: Usually two or three, similar to the brood box but
shallower.

Inner cover or **Crown board:** This may contain oval holes above which a feeder can be placed and which will take a bee-escape used to clear bees from supers.

Roof: Metal covered or well painted and ideally incorporating ventilators.

The bee-space

These basic components allow variation in the dimensions of different hives and in the provision of top or bottom bee-space. The bee-space is the gap which bees will leave clear as a passageway between the frames and the hive walls, and it is most important because without it we could not remove and replace frames as we do. The bee-space is $\frac{1}{4}$-$\frac{3}{8}$ in (6-9mm); a smaller gap would be filled by the bees with *propolis* (bee-glue) and they would build extra comb in a larger one. A bee-space must also exist between brood box and supers, otherwise the storeys would be difficult to separate.

Traditionally the British elected to have bottom bee-space, so that the tops of the frames were flush with the top edges of the box but the bottom bars were $\frac{3}{8}$ in (9mm) short of the lower edges. The cheapest kind of queen excluder is a sheet of zinc or plastic with slots cut in it, but if this is placed on a brood box with bottom bee-space the bees stick it down to the tops of every frame. Its removal then involves jolting and jarring all the frames, to the vigorously expressed disapproval of the bees. Put such an excluder on a brood box with top bee-space, that is with the frame tops $\frac{3}{8}$ in (9mm) lower than the box edges, and it will sag and distort the holes, often enough to let the queen through. The answer is a rigid wooden-framed excluder made from spaced wires, instead of stamped out of sheet, and provided with a bee-space on one side. Such an excluder is available to fit hives with top bee-space: it is imported from Germany and sold in Britain by E. H. Thorne of Wragby as the Herzog wire excluder. Since it touches only the upper

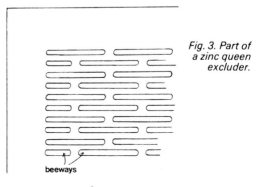

Fig. 3. Part of a zinc queen excluder.

beeways

9

Fig. 4. Part of a framed wire excluder.

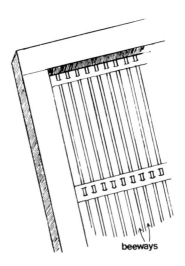

beeways

edge of the brood-box walls, it cannot be glued to anything else. Framing a zinc excluder is not successful in practice. Sheets still distort but, unlike the unframed ones, cannot be flattened out again. (The time-honoured system is to put them between sheets of newspaper under the carpet.)

Hive and frame capacity

Hives with bottom bee-space include the *National*, the most popular British hive, which takes the small British Standard frame, and the *Modified Commercial*, which takes a much larger frame. Of hives with top bee-space, one called the *Smith* takes British Standard frames. With this advantage, it is surprising that it has not replaced the National. The others are hives of American origin with much larger frames. The *Langstroth* is the hive in greatest use worldwide, being more or less standard in America and Australia. The *Jumbo*, also called the *Langstroth Deep*, has the same perimeter measurements as the Langstroth but a deeper brood box. It takes a frame the same size as that used in the largest hive of all, the *Modified Dadant* or *MD*, which accommodates eleven frames spaced at 1½in (37mm) to the Jumbo's ten at 1⅜in (35mm). This MD hive is popular with commercial beekeepers and a version of it called the Dadant-Blatt is widely used in Central Europe.

Brood-box capacity must be considered in choosing a hive. It is generally agreed that the National brood box is too small for a healthy colony with a young vigorous queen and encourages swarming due to congestion. To solve this problem, a system called 'brood and a half' is used, which means giving the queen a super as well as the brood box below the excluder. This allows a larger brood nest but means buying an extra super and providing twice as many frames and sheets of wax foundation, and also having to examine twenty-two frames instead of ten or eleven during swarm-control routines. (The short cuts recommended by some beekeepers are not reliable.) It seems more sensible to adopt a hive with a single adequate brood box. The Langstroth is just about big enough, except where very prolific bees are kept. The Jumbo or Langstroth Deep is in my view the best. It ensures that really strong colonies can be raised — and only strong colonies store surplus honey — and that adequate food can be stored in the brood box to bring the colonies through the winter. I have never found it necessary to feed colonies in these hives in the spring, which is not the case with those in smaller ones. It must be remembered that small hives can only be extended by the provision of extra equipment, whereas a large hive's capacity can be reduced by using a dummy frame. This is simply a piece of hardboard or plywood cut to the normal frame size and fitted to a top bar, so that it can be used to fill a gap or act as a false wall within the hive.

The table below summarises the number and size of the brood frames of the six hives described above.

Hive	Bee-space	Brood frames	Size of brood frames
National	Bottom	11	14 x 8½in (356 x 216mm)
Modified Commercial	Bottom	11	16 x 10in (406 x 254mm)
Smith	Top	11	14 x 8½in (356 x 216mm)
Langstroth	Top	10	17⅝ x 9⅛in (448 x 232mm)
Jumbo	Top	10	17⅝ x 11¼in (448 x 286mm)
MD	Top	11	17⅝ x 11¼in (448 x 286mm)

Frames and spacers

The light wooden frames which go inside the hive are fitted with sheets of wax embossed with the cell pattern. The bees build out cells on both sides of this midrib. By providing foundation we save the bees from having to make so much wax themselves and ensure that straight combs are built exactly within the frames. Brood frames are spaced 1⅜in (35mm) or 1½in (37mm) apart. Various ways of doing this have been devised. Clearly if frame ends touched all the way down they would be glued **immovably** together. (Manley honeyframes are made this way for convenience in uncapping but they need to be taken out only once, at the end of the season.)

*Fig. 5. Hoffman
self-spacing frames.*

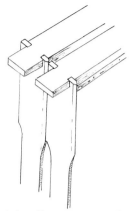

In Britain the 'metal-end' spacer was adopted, which meant that the frames had to have long lugs to accommodate it, which in turn meant that the hives had to have extra pieces of wood to form rebates deep enough to take them. Metal ends get distorted with propolis, they provide lots of little crannies inaccessible to bees in which wax-moths can pupate, and not infrequently one cuts a finger on them. In America, the Hoffman type of frame was developed. This has short lugs, easily accommodated in rebates cut in the side walls of the brood box, which is simply made from four boards. The side bars of Hoffman frames are widened at the top to form self-spacing wings and have bevelled edges to ensure that the contact area is small and propolisation minimal. There was considerable prejudice against Hoffman frames in Britain but their efficiency and simplicity are now being appreciated. Frames with straight sides can be adapted to Hoffman spacing by using Morris plastic converter clips: these are pinned to the side bars at the top, preferably on the inside to preserve the bee-space. They are useful if second-hand equipment is to be modernised, but the cost of the cheaper frames and two clips for each, in addition to pins and bother, probably offers little advantage over normal Hoffman frames.

Second-hand hives

The hives mentioned are all readily available new, but although Nationals can often be obtained second-hand, Langstroths and Jumbos seldom can. They are comparatively new in Britain, and perhaps their owners are less inclined to give up beekeeping. Most second-hand honey extractors will only take British Standard frames, so a new one must be bought to take the larger frames. Never be tempted to buy second-hand hives of different types. If you have more than one hive, all parts should be interchangeable.

Be careful about second-hand hives. Whatever type of hive it purports to be, it must be the standard article — some home-made hives have very strange dimensions. Reject hives with open joints, rot, or damaged edges so that the boxes do not fit snugly together (caused by levering apart with an unsuitable tool). Well-maintained hives have, however, a long life. It is generally wiser to discard the combs, and probably the frames too, and start afresh. The former occupants may have died from disease and microbes could infect any colony subsequently introduced. Old combs can be cut out and melted down. In any case, combs left in a hive with no bees are likely to be infested with wax-moth larvae. These white caterpillars make cobwebby tunnels through the brood combs later pupating in tough white cocoons. Remaining combs must be boiled down. Frames in good condition can be thoroughly cleaned and fitted with fresh foundation. Scrub all parts of the hive with copious quantities of hot water and soda, and even — as an extra precaution — run the flame of a blowlamp over the inside, since the germs of the worst bee diseases, American and European foulbrood, can only be destroyed by this means. Other second-hand equipment, feeders and such like, must also be thoroughly cleaned.

Making hives

Those wanting to start with new equipment can cut costs in various ways. Hives can be built by those with the necessary tools and ability. The Ministry of Agriculture leaflets giving instructions for building four modern hives have been discontinued unfortunately but you may be able to borrow the one you want. Alternatively another beekeeper might lend a hive to copy but make sure it is a well made standard example. Remember, if slightly thinner or thicker timber is used the *inside* dimensions must be preserved, or trouble with frames and bee-space will result.

Appliance manufacturers allow a discount to purchasers of hives 'in the flat' for home assembly. The parts are accurately cut and easily put together with waterproof glue (Cascamite is good) and thin galvanised nails. Assemble the boxes, square them up, then tack bits of lath across the corners to hold them square until the glue dries.

Assembling frames and foundation

Frames bought in the flat are easily assembled, providing certain points are checked before nailing. Specify top bars with wedges, and Hoffman self-spacing sidebars. Order *wired* foundation appropriate for the frame size, and fit it as you assemble the frame. Foundation is brittle when cold, so bring it into a warm room for some hours before using it. The wedge is first removed from the top bar and the whiskers of wood cleaned from the rebate with a knife. Slot the frame together ensuring that grooves in the side bars are inside. With Hoffman frames, the bevelled edge is always on the *left* facing you when the frame is the right way up. Nail through the sides into the top bar. The bottom is formed of two narrow strips. Put one in place before fitting the foundation. Some foundation has crimped wires running vertically through it at intervals; another kind has a zigzag of wire with loops top and bottom. The projections of the first and the larger loops of the second are bent to fit the angle of the rebate in the top bar, so that when the wedge is pinned back in place they are securely held. The other bottom bar is then fitted and nailed in place. The proper fine frame nails are ¾in (18mm), no. 19 gauge, gimp pins.

Super frames are simpler. Manley honeyframes have wide straight side bars, and the top and bottom bars the same width as each other so that the uncapping knife runs smoothly along them. When ordering foundation you specify 'shallow' for the relevant hive. If the beekeeper intends to eat comb honey cut from the frames, he will buy *thin unwired* super foundation. The normal thicker midrib is less pleasant to eat, though it gives strength to combs which are to be used many times.

Inner covers and clearer boards

The combined inner cover and clearer board, sometimes called a crown board, is less satisfactory than separate items. The central hole, made to take a bee-escape or give access to a feeder, causes a cold draught to rise through the middle of the bees' winter cluster, but if the hole is covered to prevent this, ventilation — essential to carry away the water vapour produced by the cluster — is impeded. It is possible to block the hole and raise the crown board with matchsticks under the corners: this gives ventilation round the perimeter as long as the bees do not propolise the gap. Another idea is to make another inner cover and keep the standard article as a clearer board. E. J. Tredwell, formerly Beekeeping Lecturer for Hampshire, advocated one with ½in (12mm) holes drilled around the edge, and sides deeper than usual, about 2½in (62mm). This facilitates the use of certain kinds of feeder (see page 41), as well as providing an insulating cushion of air and excellent ventilation above the cluster. If contact feeders will not be used, 1in (25mm) is a suitable depth for the sides. The base is marine ply, the sides at least ¾in (18mm) thick timber, glued and nailed. For use in hives with bottom bee-space, a ¼in (6mm) thick framing must be added underneath to give a bee-space over the frames. If a clearer board is also needed, being only for occasional use, this can be made with a hardboard base and a shallow wooden framing to supply bee-space.

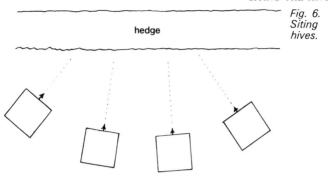

Fig. 6.
Siting
hives.

Siting the hive

The position chosen for the hive should if possible be open to the south-east, so that the sun warming one side of it calls the bees out early. The entrance should not face north or east, receiving the full force of cold winds. Avoid a hollow, which is usually a frost pocket : freezing air will drain away from higher ground. A tree or building which will shade the hive in the middle of summer days is an advantage. See that the bees' flight path does not cross a drive or road or someone else's garden. A hedge or fence about 6ft (1.8m) high and 8ft (2.4m) from the front of the hive will make the bees rise and fly over the heads of people passing or working on the other side. Do not stand the hive against a wall or hedge: leave room to get all round it. If you have two hives (and this offers certain advantages, see page 44) they can be side by side about a yard (or metre) apart. Three or four hives are better arranged in a curve rather than a straight line so that the angle of approach to each is slightly different and the bees do not confuse them.

A modern hive needs a stand to keep the floor dry. Concrete blocks are satisfactory and easily obtained. Make sure they are firm, put the hive on them and check with a spirit level that it is level from side to side. The front should however be slightly lower than the back to allow damp to run out of the entrance, not in. Slips of slate are useful for adjusting this. If the hive is on grass, keep it short so that it does not make the floor wet or obstruct the entrance.

Obtaining bees

The beginner can start beekeeping with a stock, a swarm or a nucleus.

A *stock* is a full or nearly full-sized colony already on combs with a laying queen, brood of all ages, and stores. It is an expensive way to start and a beginner lacks the experience to carry out the necessary swarm-control procedures, so it is not advisable.

15

A *swarm* is a variable number of adult bees which have left a hive with a queen but without combs and brood. A prime swarm may consist of thirty thousand bees and will have a laying queen at least one year old. Smaller swarms, which leave the parent colony after the prime swarm, are called casts and contain virgin queens. Swarms may be available without payment but there is always some risk in taking in a stray swarm. If a prime swarm is available from known bees, this is a good way to start. The bees are ready to begin waxmaking and if put on to frames of foundation they will soon draw out combs. In a good season an early swarm might give a little surplus honey the first year but it cannot be expected.

A *nucleus* is a small colony of bees on three, four or five frames with a young laying queen, brood and stores. This is a good starting point because a beginner will more easily learn to spot the queen and to examine frames in a small colony, which will develop as his experience does. It will build up to a full-sized stock by the winter, ready to store surplus honey the next season.

Whoever supplies your bees may, if he be local, help you install them in your hive; if not, an experienced beekeeper may lend a hand. The beginner's first experience of handling bees alone should not be the transfer of a nucleus nor the hiving of a swarm. However, should help not be available, the method of transferring is given in Chapter 3, and instructions for hiving a swarm will be found on page 35.

3. Starting to handle bees

Protective clothing

A *veil* is the only real essential. The simplest kind is already attached to a cotton hat with a stiffened brim to keep it off the face, or can be bought separately to wear with a hat you have. The lower edge of the veil should run over the top of the shoulders, not be pulled down over the upper arm. Various other veils and helmets are available but those with metal mesh are rather hot and heavy.

A white zipped *bee-suit* is useful. Cotton is best, though more expensive than nylon, which upsets bees which settle on it, probably due to the static electricity generated in nylon by friction. Bees also dislike animal products like wool, fur and leather, so do not tuck your trouser ends into your socks (wear wellingtons) and avoid tweed jackets and protruding woollen cuffs. It is wise to take off a watch and any ring you wear (very difficult to remove from a swollen finger after a sting).

Gloves are a vexed question. They are better not worn, especially at the start. Some beekeepers have always worn them and always will, but bee-proof gloves are thick and get stiff with dirt and propolis. They make people clumsy, which upsets the bees. Secure in his bee-proof gear, the beekeeper may be unstung, but what of innocent passers-by and neighbours? Manipulating frames with bare hands is easier and the beginner quickly learns to be neat and gentle. Gloves must anyway be discarded when clipping and marking queens. Learn without them and you will probably never want them. Instead, run elastic through the ends of the bee-suit sleeves or make cotton gauntlets with elastic at both ends so that bees do not crawl up the sleeves. If that should happen, hold up your arm and the bees will crawl up and out. They have a strong tendency to travel uphill.

Stings

Every beekeeper is stung at some time. Bees tend to be short-tempered after the main honey flow, which usually means during August in Britain, but fortunately there is no need to disturb them at this time, and once the drones are expelled and the bees begin settling down for the winter they are quite amiable again. Beginners may be uncertain in their movements and crush bees between the hive parts. The smell released by crushed bees incites other bees to sting, as does the smell of venom after one has. It is natural to jerk your hand away, but slow deliberate movements are essential if the bees are uneasy. If you are stung, scrape the sting out of your skin with your hive-tool or fingernail (it continues to pump venom into the wound even after it is detached from the bee), then mask the smell with a puff from your smoker. Do not squeeze it or scratch it. A dab of dilute Dettol is soothing and does not upset the bees. If a bee gets inside your veil, do not rip it off — there are far more outside — but puff some smoke to keep the bees quiet, walk well away and quietly release the bee. It will be as relieved as you. Once it has stung it cannot sting again and will shortly die. It is normal for a sting to swell up and cause itching and redness. Allergy is quite different, involving a rash on other parts of the body and a feeling of sickness. An allergic reaction becomes more severe with successive stings and the only course is to stop beekeeping. Normal reaction becomes less severe and eventually a degree of immunity builds up. People who go to great lengths to avoid stings never give themselves a chance to build up immunity, though nothing gives such confidence in handling bees.

Tools needed

A good *smoker* is vital. Buy a bent-nosed one with a fairly large firebox, as seen in plate 1. The small straight smoker often sold with beginners' kits is useless. Beginners take longer examining a hive and the fuel is all burnt before they have finished. Copper smokers last longer than tin ones. For fuel, old sacking or corrugated cardboard is torn into strips and rolled to fit the

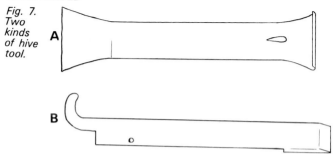

Fig. 7. Two kinds of hive tool.

firebox. Light the lower end, insert the roll partway into the smoker, work the bellows until it is going well, then push the roll down and close the nozzle. Continue steady pumping. The smoker will remain alight if stood upright but will go out if laid — or allowed to fall — flat. After use, push a plug of fresh grass into the nozzle and lay on its side. Clean out the charred remains and any coke before using it next time.

Bees fasten down any movable hive parts with propolis or bee-glue collected from trees, so a *hive-tool* is needed to lever apart the boxes and loosen frames. Choose one of the kinds illustrated. The hooked end on A acts as a scraper; the curved end on B is useful for levering out the first Hoffman frame in a closely packed hive.

Two other tools, useful but not essential, you can make yourself. *Manipulating cloths* are pieces of stout calico the same size as the top of the brood chamber, tacked or glued to a narrow lath at each end. They are used to cover the frames during examination, one being rolled up and the other unrolled as you proceed. The bees are quieter when covered, brood is less likely to be chilled on cold mornings, and a cloth is handy to cover a box temporarily, perhaps while you fetch something. Two similar strips of lath with a strip of foam plastic ¼in (6mm) thick sandwiched between makes an excellent *bee-brush*, used to brush bees from a comb or fence. A large goose feather will do the job equally well.

Opening the hive

Before approaching the hive, make sure your smoker will not go out just when you need it. Move quietly: any bumping or jarring (dropping a hive-tool on the roof, perhaps) is communicated to the bees as vibrations through their feet, causing instant alarm. Puff a little smoke in the entrance, wait a minute, then standing behind or beside the hive, carefully lift off the roof. Put it upside down to one side. Loosen the inner cover with your hive-tool, lift one corner or slew it a little and puff some smoke in the gaps. Smoke causes bees to run into the open cells and start gorging themselves with honey, making them less aggressive. You must gain control of the bees straightaway and maintain it with a gentle reminder from time to time. Once a cloud of bees has roared into the air, it is too late. Knowing when and where to smoke is learnt by experience and by watching a capable beekeeper.

The cover board is removed and stood askew on the upturned roof, which avoids hurting bees on its underside. If there is a super on the hive, it and the inner cover can be removed together. Make a gap between it and the queen excluder, puff in some smoke, and lift off. Puff smoke across the excluder to drive the bees down, lever it off, shake off the bees over the brood box, and put it aside. To shake off the bees, hold the excluder firmly with one hand and hit that wrist a sharp blow with the other hand.

There is always a little spare space in a brood box, so the next step, if you have the recommended Hoffman frames, is to lever the whole block of frames to the far side of the box with the hive-tool, so allowing as much room as possible for withdrawing the first frame. Loosen the ends of the first frame in turn, put the hive-tool somewhere handy (later you will be able to hold it as well), grasp the frame lugs with thumb and forefinger and lift it slowly up, trying to avoid rolling bees against the next comb or the hive-wall. Once it is clear, adjust your grip with your palms against the side bars. You can now study the side of the comb nearest to you.

Brood combs must never be held flat; when they are warm and heavy with brood and stores the whole comb could drop out of the frame, and honey drips everywhere. To turn the frame round, a series of movements is used which keeps the comb vertical at all times. Practise beforehand with an empty frame. Hold the lugs with the fingertips, then lower the right hand until the top bar is vertical. Revolve the frame away from you, using the top bar as the axis, so that the bottom bar is now on the right; then raise your right hand until the top bar is once again level and the frame upside down but still vertical. Adjust your grip to hold it firmly and inspect the side of the comb now towards you. To restore the comb to its former position, the same movements are carried out in reverse (see fig. 9).

Outside combs seldom contain brood. This being so, you can shake all the bees off the first frame and prop it against the side or front of the hive where you will not kick it. This gives much more room for removing and replacing other frames without damage to the bees. Each frame is examined in turn and replaced in its

Fig. 8. A home-made bee-brush.

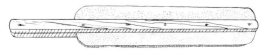

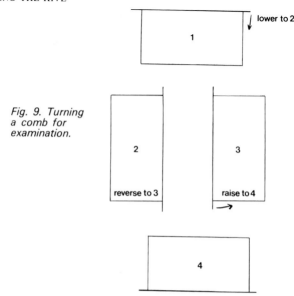

Fig. 9. Turning a comb for examination.

former position so that the exact formation of the brood nest is maintained. The frame being handled should always be held over the brood box. A laying queen is heavy and unwieldy, and if she fell from the frame into the grass she would seldom regain her hive. Also, workers will clear up any honey which drips on to the frame tops, but any dripped outside the hive attracts wasps and robbing bees. A queen is particularly liable to damage if she is on the wooden frame, rather than the comb, when it is being replaced, so check this.

A beginner examines the combs to familiarise himself with the normal appearance and development of brood, and to learn to recognise the different castes of bee, pollen, and honey. It may seem impossible to pick out a queen among so many bees but, once you have seen one, a queen is unmistakable, and facility in spotting her comes with experience. She is most easily found when the colony is small in spring, and usually on the central combs of the nest. If there is a laying queen there will be eggs, looking like tiny white dashes in the bottom of the cells. Light reflection can mislead, so tilt the comb a little each way as you search for them. Learn to distinguish capped brood and capped honey: brood will occupy the centres of the combs, honey the top outside corners. Between the two will be a ring of cells filled with pollen which may be green, yellow, orange, brick red, even black, according to

20

source. Drone cells with distinctive domed cappings tend to be in the lower corners of the outermost combs. Later, when the novice has a full-sized stock, this will be examined comb by comb in the same way for signs of swarming. Even a nucleus may have to raise a new queen if its own is injured. Queen cells are not always as obvious as might be thought, because workers often cluster thickly over them and they are often cunningly sited just inside the frame bars. At a certain stage in swarm control, when every cell must be found, the bees must be shaken from each comb in turn to allow a thorough search to be made.

When all combs have been examined, replace the first frame if it was left out, the same way round as before, lever the whole block to one side to push them close together, then centre it to leave an equal space each end. Puff smoke over the top, and replace the excluder, checking that it is the right way up if there is a difference. Put back the super and inner cover, and the roof, checking that they are exactly straight to avoid gaps. Collect the smoker, hive-tool and so on before leaving. It is worth making a box to carry your tools, including the smoker after it is cool, also spare fuel, a notebook and pencil, and later on such things as drawing pins and queen cages. Beginners can usefully make more comprehensive notes of anything seen during examination, rather than the word or two an experienced beekeeper puts on his record card. Something not understood at the time often becomes clear later, in the light of other developments. One can work out at leisure what has happened or will happen and look for the confirmation next time.

Installing a nucleus

Put the travelling box containing the frames and bees on the stand where you intend the hive to be, open the entrance and allow the bees to fly. When you are ready, prepare a feed of syrup (see page 41) and the number of additional frames needed to complete the brood box, and light your smoker.

Lift the box forward off the stand on to the ground and put the hive in its place. Give the bees a puff of smoke, and remove the lid from the box. Take out the cage containing the queen and put it in a safe place for the moment. Transfer the frames to the hive in the same order, closing them together in the centre of the brood box. Knock the nucleus box sharply on the ground to loosen the remaining bees' hold on it and tip them into the hive. A puff of smoke will send them down on to the combs. Put frames of foundation in at both sides to fill up the brood box. Tear the paper off the queen cage, make a hole through the candy stopper, then wedge the cage between the top bars of the occupied combs. Put on the inner cover, your feeder of sugar syrup, and the roof. The entrance block should be in place with the winter opening. The bees need a gallon (five litres) of syrup to settle them down and give them a good start. They will release the queen themselves and the cage may be removed a few days later.

---

4. Spring and summer

Early spring is the time to check equipment and order anything needed, before the rush. Make sure there is an excluder and a super ready in good time for each hive.

Bees fly quite early in spring if weather allows. If they are carrying pollen, this is an indication that the queen has survived and is breeding. It is normal for them to clear the winter accumulation of dead bees from the hive some fine day, but if many corpses continue appearing, send a sample for analysis (see page 46). Starvation is not uncommon in spring. Experienced beekeepers can judge the remaining stores by 'hefting' — lifting the front and back of the hive alternately — but this conveys little to a novice, unless he has two similar hives and one is noticeably lighter. He could lift the cover and look between the frames, choosing a warm day. If plenty of bees and capped honey in the tops of the combs can be seen, all is probably well. If in doubt, feed light syrup (page 41). Bees in small hives are particularly at risk.

Bees need water in spring to dilute stored honey and may be a nuisance around sink outlets and such like. Provide a supply in good time: a tin tray or plastic bowl filled with pebbles or floating sticks on which bees can settle is adequate. Put it where the bees will bother no one, where the sun can warm it a little, but not so close to the hives that young bees on playflights will soil it with excreta.

Spring examination

Choose a still warm day, probably in late March, and take a card to record findings. This examination is intended to give you the answer to certain questions:
1. How strong is the colony? This can be judged by the number of seams between combs occupied by bees. If all but the outer two are full of bees, put on an excluder and super. It is better to super too early than too late.
2. Do the outer combs contain plenty of honey? If not, feed light syrup. Do not put a feeder above a super unless it contains foundation to be drawn out.
3. Are there eggs and larvae in the brood combs? This confirms the queen's presence.
4. Is the brood in regular concentric patches with plump pearly larvae and flat even cappings? Any abnormalities need

investigation (see page 45).

5. Is there pollen in the cells around the brood?

6. Are there larvae in drone cells or capped drone brood?

7. Has the queen room to expand her brood nest? Sometimes so much stored honey remains that she is cramped. If a super is put on the bees may move the honey up into it. Sometimes it is best to remove a comb full of old pollen and substitute empty comb or foundation.

8. Can you find the queen? There is no need if eggs are present, unless you wish to mark or clip her, but practice is useful.

9. Are there many dead bees on the floor below the frames? Bees die during the winter and some colonies are more meticulous about clearing them out than others. You can lever the brood box off its floor, set it on the upturned roof and take the floor a little distance away to scrape it out. In spring, propolis is brittle and parts tend to jerk apart, so make sure the queen has not been shaken into the floor before removing it. If so, pick her up between finger and thumb around the thorax and let her run down into the brood box. Also look in the roof after replacing the brood box on the floor.

10. If you have more than one hive, are they equally advanced? If not, what is the reason?

Boosting a colony

For storing honey, a colony needs to be at its peak when there is maximum forage available: such periods are called honey-flows. In many areas the fruit blossom provides the first flow. A backward colony may build up quickly if fed with light syrup, provided no disease is present. Where there is a marked difference in the strength of two colonies, a comb of *sealed* brood can be taken from the stronger and exchanged with an empty comb in the other, first shaking all the bees from both over their own hives. This can benefit both, perhaps delaying swarming in the stronger colony by giving the queen more room to lay, and giving the weaker an influx of young bees. Add only one comb at a time as the brood will die if there are insufficient workers in the colony to cover it.

Uniting

Two weak colonies can be united, if both are healthy, but one of the queens must first be destroyed. Having removed her, spread a sheet of newspaper over the brood box and poke half a dozen slits in it with the hive-tool. On top, put the brood box containing the *queenright* stock, that is the stock with a queen, the inner cover and roof. The bees will chew away the paper and combine as one colony with the same hive scent under the one queen. Later the combs can be sorted into one box.

Marking and clipping queens

Queens may be marked for ease of finding them and for identification. Quick-drying modellers' enamel is used, a dab being applied to the top of the thorax while she is held between thumb and forefinger. A grass stem acts as a paintbrush. There is a colour code which relates to the last digit of the year of her birth: 0 or 5 is blue, 1 or 6 white, 2 or 7 yellow, 3 or 8 red, 4 or 9 green. A larger beekeeper can thus date his queens at a glance, but for a person with one or two hives white or yellow shows up best. It is sometimes hard for a novice to be certain which hive a swarm comes from if his control fails, but if he marks each queen (and the hive she inhabits) with a different colour, the swarm's origin is at once obvious.

Clipping is done as part of a swarm-control system. About two-thirds of both wings on one side are cut off with a small pointed pair of scissors, the left ones when her year of birth ends with an odd number, the right for an even number. It is all too easy to cut off a queen's leg at the same time, making her useless, so some people practise with a few expendable drones, afterwards destroying them. Care must be taken not to clip a virgin queen, who would then be unable to mate. Done early in the year there is little risk of that.

Swarming

Most bee colonies want to swarm most years since this is their natural method of increase, and they choose the time when the swarm has the best hope of survival, just before the main honey-flow. The beekeeper bent on obtaining a good surplus of honey will do his best to frustrate this urge. Congestion, which often causes preparation for swarming, can be minimised by using a large hive, and also by supering early so that the growing number of bees are not cramped into the brood box, particularly dangerous when wet weather keeps the whole colony confined all day. Giving the waxmakers something to do by substituting one or two frames of foundation for outer combs in the brood nest will also delay swarm preparations sometimes. The combs removed can be carefully stored in plastic bags and used for hiving swarms or building up a nucleus, as they will contain valuable stores.

Swarm prevention

Many highly ingenious methods of swarm control have been devised, some of which require very accurate timing or extra equipment, or both, and some — by dividing the colony artificially — result in a lessened crop. The system I have used since I began beekeeping avoids these drawbacks and seems to me less liable to failure than most. It requires that the queen be clipped and that the beekeeper opens and examines his hives once a week during

1. Beginners receiving instruction at an agricultural college. The man on the left is holding a smoker.

2. *A queen with worker bees. Note her larger size.*

3. Larvae in worker cells, four and five days after egg-laying.

4. A drone, or male honeybee. Note his heavy build and large eyes.

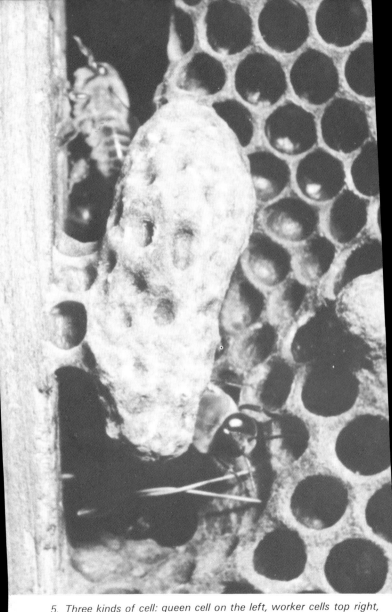

5. *Three kinds of cell: queen cell on the left, worker cells top right, and drone cells bottom right.*

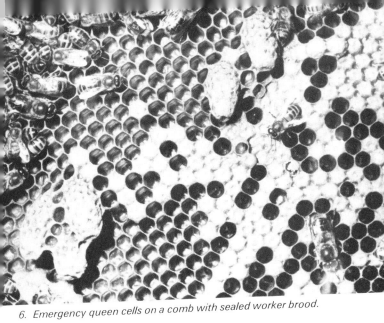

6. Emergency queen cells on a comb with sealed worker brood.

7. Worker bee collecting pollen from a wallflower.

8. Natural combs of a 'wild' colony of honeybees in a roof.

9. Worker bee on clover gathering nectar with tongue fully extended.

10. Spring examination of combs. The points to look for are described in Chapter 4.

11. Beekeeper with newly swarmed bees.

12. Swarm of bees entering a hive.

the swarming period. It can be done at nine-day intervals but it is easier to remember a set day each week, and there is then a day or two in hand if the weather is bad at the appointed time.

The combs are examined at each inspection and the presence of eggs and queen cells is checked. In practice every comb need not be scrutinised every time; queen cells do not occur on outer combs if there are none on central ones. However, once any queen cell is seen, the bees must be shaken off every comb in turn and a careful inspection made. These queen cells can be destroyed without exception, *as long as there are eggs in worker cells.* This is essential because if bees are left without a possibility of raising another queen the colony will die. Sometimes the bees give up building queen cells, but routine examinations should continue as this may be caused by weather conditions and an improvement could revive the urge.

The day will probably come when no eggs are present. This indicates that the queen has left with a swarm and, unable to fly properly, has dropped in the grass, whereupon the swarm has returned to the hive. They have no queen, so now you must leave them one open queen cell. We choose an open cell because it is less liable to damage as we handle the frames and we can make sure that it contains a living larva. Mark the frame it is on with a drawing pin or pencil and destroy all other queen cells, open and sealed. At the next inspection it would be normal to find the marked cell sealed, and several emergency cells built around young larvae. Destroy the latter if the sealed one is undamaged. No further cells can be built, and no swarm can issue if there is only one queen cell, so nothing further need be done except to check a month later that the new queen is laying. In the meantime keep away from the hive and do not alter its position or that of anything around it. The queen may make only one brief flight to learn its location before making her mating flight, and any change may lead to her confusion and loss. If you must work near the hive, do it before 11 a.m. or after 4 p.m. because mating flights take place between these hours.

A clipped queen has occasionally been known to climb the stand, re-enter her hive and begin laying again. This would be evident at the next inspection and steps will be taken accordingly. If the beekeeper has to miss an inspection, perhaps due to a holiday, he can remove the old queen, leaving one cell for her replacement, if he will be back in time to destroy any emergency cells which may be built.

This method involves no interference with colonies which do not intend to swarm and makes clear which have a lesser tendency to do so. The colony is kept intact throughout the year, and regular examination ensures that an outbreak of disease cannot be overlooked. Though a beginner will take longer, experience will soon cut down the time needed to inspect each hive to about fifteen minutes.

Dealing with a swarm

Whatever measures are taken to control swarming, every beekeeper has to deal with a swarm sometime, his own or a stray one. Many people have begun beekeeping with a swarm reported to the police by an alarmed householder. Although there is some risk in taking in a strange swarm, beekeepers have some responsibility to the public and from their own point of view wild colonies (plate 8) are a nuisance as they compete with hive bees for local forage.

A swarm swirling in the air may look dangerous but it is really harmless. The bees are still good-tempered for a while after they have clustered (plate 11), but if the scouts fail to find a new home they will gradually become crosser as their reserves are used up. For this reason, as well as the chance that they will take up residence in a roof or other awkward place, they should be dealt with promptly. You need a straw skep or medium-sized cardboard carton (or two) and probably also a bee-brush, queen cage, length of wire, a sack or piece of sheet and some strong string. I like to spread a sheet on the ground in a shady place near the swarm first. If a swarm is hanging in a neat cluster on a branch, hold the skep or box firmly and close up under it, then give the branch a sharp knock or shake. The cluster drops into the skep, which is then turned upside down on the sheet, with a stone under one edge so that flying bees may join the rest. If the queen is not in the skep (and virgin queens are often elusive) the bees will find her and cluster again, and the process must be repeated.

If the swarm is spread over several little branches of a bush, secateurs may be used to cut the stems as you gather them into the other hand, and then the whole bunch is shaken into the skep. Bees may cluster around a post or tree trunk. In this case hold the skep against the tree, brush in as many bees as possible with two or three strokes and invert the skep. If the bees begin fanning their wings around the opening, with their heads down and the Nassenoff (scent) gland exposed — it shows as a white mark towards the end of the abdomen — you have the queen. Continue brushing bees into a second skep or box and shaking them on to the ground in front of the first one. The skep may be left as it is, once the queen is inside, and the bees hived in the evening after the flying ones have gone in.

Sometimes, if the scouts have located a suitable home, the swarm will fly away before the evening. Finding and caging the queen will prevent this. Look for her first on the cluster; then in the skep, rolling it gently in your hands. If this fails, throw the bees out on to the sheet (where they will show up well), put the empty skep propped up at the edge of the pool of bees and watch as they flow towards it. The queen can usually be spotted on the way. The best queen cage is a metal or plastic hair-roller with a cork in each end. Fasten it in the top of the skep with a bit of wire pushed through both.

If there is no natural shade, protect the skep with leafy branches and suchlike; and never leave a skep standing flat on the ground for more than a few minutes or the bees will suffocate. In the evening, take away the stone, gather the corners of the sheet and knot them together, then tie string around the skep to confine the bees for transport. If you have a nucleus box, a swarm can be run straight into it, the queen's cage wedged between the top bars of two frames if she is caged, and the entrance closed with a strip of perforated zinc and drawing pins for transport in the evening. The frames can be transferred to a hive next day, but reopen the entrance as soon as possible.

Hiving swarms

You need a piece of ply or hardboard as wide as the hive and long enough to form a gentle slope between the ground and the entrance when the hive is on its stand. Prop it up so that it is firm and there is no gap at the top. Uncover the skep, remove the queen if she is caged, knock the skep on the ground to loosen the bees' hold and shake them all out on to the board (plate 12). This is the time to release a caged queen, close to the entrance. A swarm, especially if it has foundation to draw out, needs a gallon (about five litres) of syrup to give it a good start.

If the swarm you have came from your own hive, two courses are open to you, depending on whether or not you want to increase your stocks (or have a spare hive). If you put it in a new hive, your honey crop will be lessened, but by moving the parent hive and putting the swarm on the old stand, you improve your chance of a

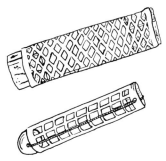

Fig. 10. Queen cages made from metal and plastic hair-rollers.

surplus and prevent further swarms issuing.

Put the roof of the new hive beside the old one and stack on it the supers and excluder from the old hive, covering them temporarily with a cloth or clearer board. Find a good queen cell, preferably open and containing a healthy-looking larva, and mark the frame. Destroy all the other queen cells. Carry this old hive to a new stand, at a slightly different angle from its former position. The new hive is now put on the old stand, the queen excluder and supers are added to it, with its inner cover and roof, and the swarm is run in. The bees are of the same colony so will not fight with those in the supers. Having stores, they will not need feeding. Foragers flying from the parent hive will join the new one, leaving mainly young bees to rear the brood and raise the new queen in the parent colony. If combs rather than foundation are available, an augmented swarm like this will often store a reasonable surplus under favourable conditions. There will be no new adult bees for three weeks, even if the queen has somewhere to begin laying at once, so a swarm tends to get smaller before it starts to increase, since bees are dying every day.

If a colony has swarmed and no hive is available to put them in, they can be returned to the one they came from. First you must go through the parent hive and remove all queen cells except one, then the queen in the swarm must be found and killed. After that the swarm can be run in again. As long as the bees have plenty of room there is every chance they will not swarm again because the gap in brood production resembles that caused by the loss of a swarm. Swarms cannot be run into a strange hive, though they can be united by the newspaper method if one of the queens is killed.

Artificial swarms

A beginner may be tempted to let his first colony swarm in order to obtain a second, but apart from alarming the neighbours there is a risk of losing the bees. It is better to make an artificial swarm.

When you find queen cells in the hive, you move it off its stand and put a new hive in its place. Take out the two central frames from the new brood box and put aside. Find the comb with the queen on it in the old hive, remove any queen cells which may be on it too (without shaking off the bees) and put it in the space in the new hive. In the remaining space put a comb of stores, honey and pollen, from the old hive, first checking that there is no brood or queen cell on it. Now shake the bees from two other combs into the new hive, but replace the combs in the old one. The new hive now contains the 'swarm', consisting of old queen, young bees and some stores. Most of the foragers from the parent stock will join it on the old site, so you put the queen excluder and supers from the old hive on it and close it.

Choose a comb in the old hive with a good queen cell on it and mark the frame. Leave it in the hive, and destroy all the other queen cells. Fill up this brood box with the two frames taken from the new hive, and you now have a colony as it would be after swarming. Put it on a fresh stand, facing in a slightly different direction. A few days later check for emergency queen cells and destroy any found. Some people prefer to wait until the day after making the artificial swarm before dealing with the queen cells in the parent hive because the flying bees will by then have joined the swarm (weather permitting) and the hive will be far less populous.

5. Autumn and winter

At the end of the swarming season the beekeeper can relax. If they have adequate super space, bees should not need attention during August, and the next task will be taking the honey.

Cost of equipment

Equipment for extracting honey is expensive for the person with one or two hives. Some associations have an extractor for hire, or two or three beekeepers might buy one between them. An alternative is to work only for comb honey. The small square sections are difficult for beginners as the bees dislike them and tend to swarm, and only a proportion of those put on the hive are actually completed. Honeycombs produced in normal super frames fitted with thin foundation are sold as 'cut comb', divided into blocks to fit small plastic boxes. The home consumer could simply scrape the honey off the midrib (using normal foundation), returning this for the bees to clean up and to build new comb on the following year.

The extractor

For run honey an extractor is needed, of the correct size to accommodate the frames used. It is a straight-sided metal or plastic tank containing plastic-coated racks to hold the frames, which are whirled round by turning a handle. Honey is thrown out by centrifugal force and runs down to be drawn off by a tap at the base. Electric models are available for the larger concern.

Uncapping tray

Before placing in the extractor, cappings are sliced off the combs. The frame is stood on end, slanting forward so that, as the knife cuts upwards from the lower edge, the cappings fall cleanly into a tray below. The heated Pratley tray separates wax from honey and is convenient, but a plastic washing-up bowl or bin can be used. If a mesh can be fitted a few inches above the bottom, much of the honey will drain from the cappings as you work. Have

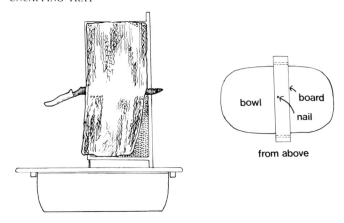

Fig. 11. Uncapping honeycombs.

a board to rest across the top. A nail hammered upwards through it to the length of a frame lug holds the frame level and firm during uncapping.

Uncapping knives

There are scallop-edged uncapping knives which are used cold and are said to be efficient. Electrically heated ones are good (and expensive) but the job can be done with two carving knives to use alternately, kept hot in water in a tall tin or old coffee-pot. Put a clean cloth beside it and give each knife a quick wipe before use, or knock off the water by giving the back of the knife a sharp tap on the edge of the tin, so that you do not get water in the honey. Wipe the honey off the knife blade on the edge of your uncapping bowl before replacing it in the hot water.

Straining tank or ripener

The honey is ripe when the bees cap it. The ripener is simply a tank containing strainers with a tap in the base, into which honey is run from the extractor. A plastic fermentation bin can be used with a tap of at least 1¼in (32mm) bore fitted to it, or honey can go straight into storage containers through a conical strainer hung on the extractor tap. Degree of straining is a matter of taste. A conical metal-mesh strainer will remove wax particles and bits of propolis; lined with scrim or cheesecloth it will filter out the larger pollen grains. Honey always contains some pollen, which contributes much of the flavour and vitamin content and all the protein, and, as with other foods, the more refined it is the less

*Fig. 12. Conical
metal strainer.*

valuable. The honey I sell is always more thoroughly strained than that which I eat myself, because it is clearer and customers judge it only by appearance.

Containers

Honey absorbs moisture from the air, so it must be kept in airtight containers. If you intend selling it, it must be packed and labelled in accordance with the regulations. Make sure any container used is scrupulously clean, dry and odourless, and avoid part-filling large containers. If you store it in 7lb tins, decant the whole tinful into smaller jars to use. All untreated honey granulates or crystallises in time but can be reliquefied by warming. Stand the container in the airing cupboard over the hot tank or in the oven (switched off) after the Sunday joint is removed.

Taking honey

Honey is taken once a year, usually in September. Bees may starve if it is taken earlier and then bad weather occurs. It must also be ripe and sealed. The Porter bee-escape is fitted into the clearer board with the hole uppermost, after checking the gap between the prongs. The bees pass down through the hole and out between the flexible prongs, but cannot return the same way. The

beeway

*Fig. 13. Porter
bee-escape.*

board is best put under the super in the evening so that the bees clear overnight. Have a square of cardboard or newspaper to put under the inner cover if the roof has large ventilators, or bees may waste time looking for a way out where they see light. Unprotected honey causes trouble from robbing bees at this time of year, so make sure supers and roof are replaced accurately, mesh ventilators are intact, and supers are not uncovered longer than necessary. Lift them off early or late in the day, when fewer bees are about, and carry them indoors. Warm combs are easier to uncap and extract. If supers must be kept a while, make sure they are bee-proof — bees can find their way in to honey through keyholes and ventilator slots as well as doors and windows and will bring thousands more.

Check that everything you need is clean and dry, and shut the extractor tap. Uncap enough combs to fill the extractor, putting each in as it is done. Balance the load with heavy combs opposite each other to minimise vibration. Turn the handle gently until honey spatters out on to the sides of the canister. Partially extract the first side, turn the combs and empty the second side, then reverse again to finish the first; this puts less strain on the combs. Replace emptied combs in the super and uncap a fresh batch. Supers are later put back on the hives for the bees to clean off what honey remains and repair damaged cells, leaving clean dry combs for storage. Do this late in the evening to avoid disturbance and robbing. Honey strained from the cappings may be added to the rest. No honey should be left uncovered long as it absorbs moisture from the air, which undoes the bees' good work and makes it liable to ferment. It runs easily through a strainer when warm, but do not leave it unattended. Every beekeeper has a tale to tell of a blocked filter or a strainer filling faster than it emptied and the resulting silent sticky overflow.

After use the extractor must be dismantled, washed, and dried with the tap left open; then bare metal parts must be wiped with medicinal liquid paraffin and the whole thing put in a large plastic bag — and back in its box if you have it. Besides deteriorating, a rusty extractor spoils honey. Wax from cappings and strainers should be washed in warm water, strained off and dried. Wax is valuable. The water will contain quite a lot of honey and can be used as a basis for mead and other home-made wines.

Feeding bees

Bees need feeding with syrup after the honey is taken, so that they can survive the winter, but a colony may need feeding at other times and it is convenient to deal with them together. In general we feed bees to help them build combs and develop quickly, as with a swarm or a nucleus; to help them overcome disease; because they are short of stores in spring, owing to inadequate autumn feeding or to too small a hive (in mild winters bees fly more and use more food than in cold ones); because unusual weather occurring when the hive is crammed with bees means rapid depletion of stores; and in autumn.

A colony's stores can be supplemented with a comb of honey from another healthy colony, but liquid honey given in a feeder causes uproar and robbing. Honey from an unknown source, especially foreign honey, is always dangerous as it may contain germs of disease harmless to humans but fatal to bees. The only safe food is syrup made from white sugar; less refined brown sugar cannot be digested and causes dysentery. Thick syrup is made by stirring 2lb of sugar into 1 pint of water (2kg to 1 litre is not an exact metric conversion of this, but is the standard strong syrup used on the European continent). The mixture is brought to the boil and allowed to cool. This is fed in autumn because there is less water to be evaporated as the bees process it for storage. Thin syrup, made with half the quantity of sugar, is fed at other times when the bees need it for immediate use, not to store. Thin syrup may ferment if made in advance, but thick syrup can be kept in plastic cans and diluted at need. Bees can deal with large amounts of syrup very expeditiously and small feeders waste time.

Contact feeders

The simplest feeder is a large lever-lid tin such as those used for honey or instant coffee. A dozen small holes are knocked in the lid with the point of a fine nail, the tin is filled, the lid firmly replaced, and the whole thing inverted over the feed hole in the crown board. If an inner cover with perimeter holes is used, the tin can rest on two spaced laths of ¼in (6mm) thickness to allow the bees access. A sophisticated version of this is the bucket-shaped white plastic contact feeder with wire mesh let into the lid. Contact feeders need to be enclosed inside an empty hive box, but small beekeepers seldom have one, and supers which can be emptied temporarily are not deep enough. In this respect Captain Tredwell's deep-sided inner cover has the advantage, as with a super it gives the required depth. If combs are taken from supers during feeding they should be kept safe in plastic bags and replaced for winter storage.

Rapid feeders

These are round metal feeders with a hollow central pillar by which bees ascend to the surface of the syrup, and over the pillar a glass box which prevents bees floating out of reach of it and drowning. Liquid flows under the edges of the box to maintain the level. A lid covers the whole thing, although when this feeder is set over a feed hole there is no chance of bees getting into the main body of the syrup. The new plastic version of this feeder gives bees insufficient foothold unless the pillar is roughened with sandpaper or covered with metal mesh.

Overall feeders

Miller and Ashforth feeders are both wooden trays which fit the top of the hive like another storey and allow bees access to the syrup either in the centre or at one end. Though expensive, they do away with the need for extra hive boxes, can be filled without removal from the hive (as can Rapid feeders) and hold a lot of syrup. They need no cover except the normal bee-proof roof. A capable do-it-yourselfer could probably borrow and copy one.

Whatever feeder is used, check it for leaks, and avoid spilling syrup outside the hive, laying trails of it, or leaving cans and buckets about. Winter entrances should be in place before commencing autumn feeding. It is impossible to make rules about quantity as much depends on the hive in use, size of colony, weather, and forage available. Bees use some of the syrup to feed themselves as they carry, process and seal the rest. Some colonies are encouraged to increase breeding and use up syrup that way. At the same time nectar may be coming in from ivy as late as November in warmer counties. Dark native or near-native bees are less prolific and hardier than yellow-banded Italian bees (though few of either kind can be called pure-bred). Dark bees tend to curtail breeding, even at midsummer, if conditions are unfavourable, while yellow bees, from warmer climates, will continue breeding and starve. A beekeeper coming to their aid will have a powerful force of foragers ready if conditions improve, but will have had extra trouble and expense for any additional honey he may get.

Winter work

After autumn feeding is finished, while the weather is still warm enough for the bees to concentrate and store the syrup, clean and put away feeders and spare hive boxes. Remove queen excluders, make sure the brood combs are pushed tight together, check winter entrances, and close the hives, clearing away any long grass around them which encourages damp. I like to paint each roof in turn before the onset of winter, which means having a spare to put on instead.

Spare brood combs are very susceptible to wax-moth attack and are best kept in plastic bags with a few PDB moth-repellent crystals, but combs must be thoroughly aired for several days before being given to bees again. Supers should be stored in a pile standing on a flat floor and covered with a spare roof, or a clearer board with a piece of wood or glass over the hole. If a colony has died out, fumigate the whole hive as explained on page 46.

During the winter, work through your equipment, scraping wax and propolis from woodwork, and carrying out repairs to damaged frames, weak joints and so on. Decoke the smoker and order anything you will need next season before prices go up.

Wax

Oddments of wax, cappings and misshapen combs can be gently melted in an old saucepan with rainwater, and the clean cake of wax lifted off the top when the water cools. Small beekeepers would not have enough use for the wax-reclaiming equipment sold by appliance firms, but a home-made solar wax extractor deals with all waste wax cheaply and without trouble. It operates whenever the sun shines and avoids mess in the kitchen. The basis is a large roasting tin and a loaf tin which are fitted into a wooden

Fig. 14. Simple home-made solar wax extractor.

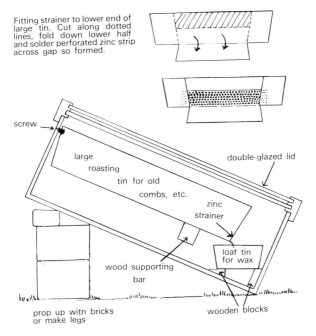

Fitting strainer to lower end of large tin. Cut along dotted lines, fold down lower half and solder perforated zinc strip across gap so formed.

screw

large roasting tin for old combs, etc.

double-glazed lid

zinc strainer

loaf tin for wax

wood supporting bar

prop up with bricks or make legs

wooden blocks

box (painted black outside and white inside) in such a way that wax put into the large tin is melted by the sun's rays striking through the double-glazed lid and runs through a strainer into the smaller tin, where it sets as it cools. The dross remains behind. Bits of comb can be usefully put inside an old nylon stocking or strainer cloth before putting in the wax extractor: this is then burnt after the wax has run through. The wax block, if not perfectly clean, can be put back for a second melting. Such blocks can be kept without deteriorating until you have sufficient to trade in for new foundation.

6. Things which may go wrong

Bees are healthy, well-organised creatures, but on occasion their arrangements break down and they can contract various diseases. The beekeeper needs to be able to recognise these conditions and provide help.

Queenlessness

This means that a colony is left without a queen or any means of raising one. The one queen cell left by the beekeeper may have been defective, or a virgin may not have returned from her mating flight. If queenlessness is suspected, take a comb containing some worker eggs from another hive, shake all the bees off, and exchange it with an empty comb from the queenless brood nest. If the bees have no queen, they will immediately build emergency queen cells. Sometimes novices imagine a hive is queenless when the queen is only taking a rest from breeding in the autumn, but no harm will come from giving them a comb of eggs as would if attempts were made to introduce a new queen.

Laying workers

A number of drones present in a hive unusually late in the year may indicate that a queen has died and the workers are raising a new one and hoping to get her mated at the eleventh hour. Where there is a large number of rather undersized drones, the problem may be laying workers. If a hive is queenless for some time, lack of queen substance allows the workers' ovaries to develop and some will lay eggs. The brood pattern is patchy with empty cells and others containing two or three eggs, and domed cappings will be seen on worker cells. As all eggs will develop into drones, the colony is doomed. It cannot be requeened. The only possibility is to shake all the bees on to a hiving board in front of another hive: the laying workers would be killed as they enter it.

Drone breeder

An old queen whose supply of sperm is exhausted will lay only infertile eggs. The brood pattern is regular with one egg to a cell but — as with laying workers — there are domed drone caps on the worker cells. The queen must be removed and the colony given a comb of eggs from another hive so that they can rear another queen.

Robbing

Preventing robbing is easier than stopping it when it has started. This entails a high standard of apiary hygiene, putting on feeders late in the evening, and restricting entrances. Weak colonies are attacked, so deal with disease promptly and keep colonies as even as possible by uniting, or building up with brood from another hive. Queenless bees are vulnerable, and indeed invite the attentions of robbers from a queenright colony. If signs of robbing are seen, restrict the entrance with a strip of foam plastic to one bee passage which the guards can defend. If the colony is killed (which can be confirmed by examining it at twilight when the robbers have gone home), do not remove the hive, or at least leave something for the robbers to clean out, because sudden frustration will lead them to attack other hives. Take the hive away when they lose interest and fumigate it, or the combs will be destroyed by wax-moths. Wax-moth eggs are hard to kill, so store the combs with PDB crystals and watch out for trouble.

Poisoning

Pesticides have proliferated in recent years and most of them are lethal to bees. Try educating your neighbours, as most gardeners are ignorant rather than ill-intentioned. The rule is *never* to spray open blossom. Sprays also drift on to other things, such as dandelions under fruit trees (cut them down first), and powder insecticides are sometimes carried into the hives like pollen and fed to larvae. Some sprays are safe if used after the bees have retired in the evening. Those based on derris and pyrethrum are the least toxic to bees. However, the greatest damage to bee colonies occurs from large-scale aerial spraying of field crops like oilseed rape. Even if beekeepers have prior notice, shutting bees up is as hazardous as letting them fly since they are liable to suffocate. There is a code of practice for spraying but a greater degree of co-operation is necessary between farmers and beekeepers. Their interests are not incompatible, and beekeepers should make themselves known to farmers in their area and try to come to an amicable arrangement.

Diseases of bee brood

Signs of brood disease include dead larvae in cells, a patchy brood pattern, and cappings which are sunken, dark or perforated. The diseases called American and European foulbrood must by law be reported to the Ministry of Agriculture. If symptoms are noticed, restrict the entrance to prevent stronger colonies robbing and thereby spreading infection (close up entirely if the colony is dead) and contact the local Ministry Bee Officer. If it is foulbrood, he will do what is necessary, destruction of bees and combs and scorching out the hive in the case of AFB, possibly antibiotic treatment for EFB. Insurance (very reasonable) is obtainable through beekeepers' associations. The trouble may be sacbrood or, more easily recognised, chalkbrood, in which dead larvae resembles knobs of white chalk. No cures are known for these diseases but correcting adverse conditions such as damp hives, a feed of syrup, or requeening might be advised. A third notifiable disease is varroa, a parasite which infests (mainly) larvae. It must not be confused with the tiny wingless fly *Braula coeca,* which may be seen clinging to adult bees, especially queens. These are fairly harmless. For recognition and treatment of brood diseases and varroa, the illustrated leaflets available free of charge from MAFF should be studied.

Adult bee diseases

The signs are large quantities of dead bees inside or outside the hive, or a number of bees crawling about unable to fly. Even experienced beekeepers cannot easily diagnose the trouble. Microscopic examination is needed for diseases such as nosema, acarine and amoeba, and this service may be available at a county agricultural college. If not, the National Bee Unit, MAFF (see appendix), will help. A sample of thirty bees should be sent in a matchbox, not a tin, glass or plastic container. Enclose a note of explanation, your name and address and the hive number. A charge is made for this service, which includes a full report and advisory notes.

Sterilising equipment

If a colony dies for any reason (other than AFB or EFB which need a different treatment) it is sensible to fumigate the combs, which can then be used to hive swarms and so on. For this you need 80 per cent glacial acetic acid obtainable from a chemist. It is a strong acid so take care and wash any splashes from the skin immediately. Carry out the operation outside or in an open shed. Put in the entrance block of the hive and cover any poor joins with paper tape. A 6in (15cm) square of cotton-wool soaked in acetic acid is placed above the brood combs and another above the supers. Over that put an inverted deep inner cover or an empty super, a sheet of polythene if the roof is ventilated, and the roof. Alternatively the whole hive may be assembled inside a large polythene bag which is then sealed over the top. Leave this undisturbed for a week or so; if longer, the acid will corrode metal parts like frame nails and wire. If metal-end spacers are used, these should be removed before fumigation and boiled in water with washing soda to clean. Combs should be aired for at least forty-eight hours before being used again.

Appendix

Further reading

Brown, Ron. *Beeswax*. BBNO, new edition 1995.
Brown, Ron. *Great Masters of Beekeeping*. BBNO, 1994.
Butler, Colin G. *World of the Honeybee*. Collins, 1971.
Dadant. *The Hive and the Honey Bee*. Dadant, USA, new edition 1992.
Dade, H. A. *Anatomy and Dissection of the Honeybee*. IBRA, 1978.
Howes, F. N. *Plants and Beekeeping*. Faber, 1979.
Matheson, A. *Living with Varroa*. IBRA, 1993.
More, Daphne. *The Bee Book*. David & Charles, 1976.
Ribbands, C. R. *The Behaviour and Social Life of Honeybees*. IBRA, 1953.
Riches, H. A. *Handbook of Beekeeping*. NBB, 1992.
Riches, H. A. *Honey Marketing*. BBNO, 1989.
Wedmore, E. B. *Manual of Beekeeping*. BBNO, 1975.
Winston, M. *The Biology of the Honeybee*. Harvard, 1987.
Yates. *Yates Beekeeping Study Notes for the BBKA Examinations*. BBNO, 1991.

Journals

Bee Craft (monthly): The Secretary, 15 West Way, Copthorne Bank, Crawley, West Sussex RH10 3QS.
British Bee Journal (monthly): 46 Queen Street, Geddington, Kettering, Northamptonshire NN14 1AZ.
Beekeepers Quarterly: Northern Bee Books, Scout Bottom Farm, Mytholmroyd, Hebden Bridge, West Yorkshire HX7 5JS.

Bee books by mail order

Bee Books New and Old, Tappingwall Farm, Burrowbridge, Bridgwater, Somerset TA7 0RY.
Northern Bee Books, address above.

Ministry of Agriculture, Fisheries and Food (MAFF)

Ministry of Agriculture Bee Officer (see 'Government departments, local offices' in Yellow Pages).
Central Science Laboratory, National Bee Unit, MAFF, Luddington, Stratford-upon-Avon, Warwickshire CV37 9SJ. Telephone: 01789 750601.

Appliance manufacturers

E. H. Thorne Ltd, Beehive Works, Wragby, Lincolnshire LN3 5LA.
Steele & Brodie Ltd, Beehive Works, 25 Kilmany Road, Wormit, Newport-on-Tay, Fife DD6 8PG.

Other useful addresses

The British Beekeeper's Association: General Secretary, BBKA Headquarters, National Agricultural Centre, Stoneleigh, Kenilworth, Warwickshire CV8 2LZ.
The International Bee Research Association: Secretary, 18 North Road, Cardiff CF1 3DY.

Index